Whisper in the Woods shares the beauty and value of nature through a carefully selected showcase of art, photography and literature. Each issue offers new ways for all of us to *Discover, Explore* and *Appreciate* the natural world we cherish.

Editor/Publisher
Kimberli A. Bindschatel

Founding Designer
Suzanne Conant

Illustrator
Rod Lawrence

Inspiration
Mother Nature

Associate Editors
Denise Baker
Pam McCormick

Contributors
Eric Alan
Kyle Bagnall
Denise Baker
Jim Brandenburg
Sreve Brimm
Mark S. Carlson
Lisa Daly
Erin Fanning
Jan Ferris
Mark Graf
Shawn Grose
Janea Little
Michael Rathbun
Roger Smith
Jean Strelka

Webmaster
Stacy Niedzwiecki

Journal Dogs
Kloe & Tucker

Special Thanks
Ken Bindschatel
Patrick Cannon

Cover Photograph
© Jim Brandenburg

Printed in the U.S.A.
Millbrook Printing
Grand Ledge, Michigan

Volume Five, No. Three

To Subscribe:
For current rates, please visit
www.WhisperintheWoods.com
or write to:
Whisper in the Woods Subscriptions
P.O. Box 1014
Traverse City, MI 49685
(866) 943-0153

Change of Address:
The post office will not automatically forward *Whisper in the Woods* when you move. To ensure continuous service, please notify us at least six weeks before moving. Send your new address and subscription number to: *Whisper in the Woods* Subscriptions, P.O. Box 1014, Traverse City, MI 49658.

Submitting art, photography or writing:
Please download our guidelines at www.WhisperintheWoods.com

We are proud to be affiliated with:

Kalamazoo Nature Center
Kalamazoo, Michigan
www.naturecenter.org

Chippewa Nature Center
Midland, Michigan
www.chippewanaturecenter.org

Nature Center at Shaker Lakes
Cleveland, Ohio
www.shakerlakes.org

Ad Sales:
Toll-Free (866) 943-0153
ads@whisperinthewoods.com

Whisper in the Woods®
(ISSN 1543-8821) is published quarterly by Turning Leaf Productions, LLC. We strive for accuracy in the articles and honesty in advertising. We reserve the right to refuse any advertising that is inappropriate or not in harmony with the editorial policy. Please obtain written permission before reproducing any part of this publication. ©2006 All rights reserved. Registered trademark.

Member, International Regional Magazine Association

Schlitz Audubon Nature Center
Bayside, Wisconsin
www.sanc.org

Cincinnati Nature Center
Cincinnati, Ohio
www.Cincynature.org

About this issue: (ISBN 0-9785820-1-2)

The wolf is the epitome of what is wild. And in the heart of our beautiful Lake Superior lies Isle Royale, a rare treasure of true wilderness, protected in its natural beauty. If you haven't yet enjoyed exploring the Isle, be sure to put it on your life list! It is my honor to publish the images of my idol, Mr. Jim Brandenburg.
~Kimberli Bindschatel, Editor

(Editor's note: All wolf photos featured in this issue were not necessarily taken on Isle Royale.)

Table of Contents

Wolves of Isle Royale

Discover

Writer
Michael Rathbun
23

  6 GORP
15 Field Notes
57 Backyard Birds
61 Resource Guide

Explore

Photographer
Jim Brandenburg
27

11 Northern Outings
19 Lasting Impressions

Appreciate

Artist
Roger Smith
49

  5 My Thoughts
47 Wild Grace
60 Poetry

Next Issue: Snowshoe Trekking

Don't miss our Winter issue
featuring our Annual Reader's Showcase

Poetry • Illustrations • Essays • Photography

Nature Journal

Check out our web site for entry rules and deadlines
www.WhisperintheWoods.com

Wolves of Isle Royale
my thoughts...

Ah, the wolf. No other animal of our north woods evokes so much emotion. Wolves are a symbol of what it means to be wild, and they hold such a sense of mystery and intrigue. I have only seen a wild wolf once. It ran quickly across the road in front of me in Ontario's Algonquin Provincial Park. Like Isle Royale, Algonquin was created as a sanctuary for wildlife; wolves thrive there, as they once did for thousands of years throughout our Eastern forests.

My experience of seeing a wolf actually left me somewhat disappointed. Like many wildlife sightings, our encounter was much too brief. We were both on the move—I, in a car, and the wolf, on four feet. I didn't get a chance to see its face.

What is that primitive excitement we feel when we catch a glimpse of a wolf? Wolves are known as killers of the forest, vicious, without remorse. They evoke in us an instinctual flight reaction as we realize we're a step lower on the food chain. The wolf, without doubt, is aware of our presence in the forest. Its sense of sight and smell are vastly superior to our own. The wolf is a hunter and a survivor.

However, if we look closer, we see the wolf is similar to us in so many ways. Each pack is a family, with every member filling a unique place in the whole. An alpha pair takes the lead; others follow, until they mature and eventually assume leadership. Packs are also incredibly efficient hunting groups, able to take on almost any challenge. Like wolves, people are at their best when they work together.

While I have yet to get a good look, I have heard wolves howling at night. The moon was full, and the whole experience felt so surreal, primitive and raw. I imagined the alpha calling the pack together to go out for a hunt. My heart pounded as wolves gathered unseen in the forest. If I ever have the opportunity to hike Isle Royale, more than anything, I look forward to experiencing the feeling of looking into the eyes of a wolf.

Photo by Mark Graf
www.GrafPhoto.com

Kimberli A. Bindschatel
Kimberli A. Bindschatel
Editor in Chief

GORP

Stargazing
through a child's eyes

No matter how many times we look at familiar sights, we don't always *see* them. Seeing and looking are two different activities. Looking is a visual activity, but seeing something involves cognition, emotion and experience. This is true for us in all moments of our lives: gazing upon our childrens' faces, navigating our familiar routes about town, or beholding our natural environment.

And it is especially true of our night sky. I often hear the comment from planetarium visitors: "I never noticed that before." It's always a joy for me to show people something that is right there in front of them; they just need a little guidance in the seeing department.

Looking at the stars, there are some tricks we can employ to see better in the dark. Allowing our eyes to adjust to darkness is one. Our pupils must dilate to allow in more light, and a photochemical reaction takes place, allowing our eyes to work more effectively. Waiting a full ten minutes is one way of learning to see better.

Learning to use a red filtered flashlight is another. Humans have acute vision. We see better in terms of color, detail and clarity than just about any animal—at least in the daytime. We all know that owls have us beat when it comes to night vision. But in the evolution of the owl's fabulous night-adapted eye, compromises were made. Owls' eyes are relatively huge in their heads to accommodate the large number of rod cells that allow them to see better at night. Bigger eyes took up room for the muscles that would allow an owl to move its eyes within its head; owls cannot look left and right without moving their heads. Of course, they can turn that head around 270 degrees!

We have more cone cells for seeing better in the daytime. Our rod cells are located a little off to the side of our visual field. When you are stargazing, make use of this; when trying to resolve a dim object, such as a distant nebula or galaxy, look slightly away from it, and it will come into view. By relaxing your eye, and using your night vision, you will see more of the sky than you have been looking at.

I hope that I have opened your eyes for you as a kindergarten visitor to the planetarium once did for me. Upon seeing the Milky Way galaxy brilliantly stretched across the planetarium sky, he proclaimed it "a rainbow that has lost all its colors!" This image persists for me whenever I seek our beautiful galaxy. He opened his eyes and his mind, and allowed me to see.

–Lisa Daly

GORP

Quiet Sports
the perfect sole: how to buy great hiking boots

My ankles wobbled and my soles ached as I dumped my backpack near the tent and plopped down next to it. Leaning against the heavy pack, I rotated my ankles, which easily flexed in my soft hiking boots that were worn with age. Still standing after our climb of several miles and a few thousand feet in elevation gain, my husband Keith, sturdy leather boots hiding his feet, flashed me a told-you-so smile. I grimaced and continued massaging.

Like most sporting equipment, hiking boots come in many flavors—from lightweight day hikers, like the ones I had mistakenly chosen for our backpacking trip, to mountaineering boots, ready for crampons and days of climbing with a laden pack. Within these two extremes fall a variety of gradations and countless brands, offering a model or two in each type of boot. The key is to winnow your options based on the boot's intended use and your own hiking experience.

First, you should decide where and how you plan to use the boots: short day hikes, over even ground with an occasional hill, or longer trips in mountainous terrain? Will you carry a heavy backpack? And what can you expect from the weather: calm summer skies or a riotous late fall day with snow flurries? Will you encounter slippery rocks and other slick obstacles?

Second, you should consider your own hiking experience. More experienced hikers can sometimes get by with a lighter, softer shoe than the less experienced, who may need the extra support of a more beefy boot.

Once you've decided how you plan to use the boots, visit a reputable sports or outdoor store. Although hiking boots are sold at a variety of department and big-box stores, you typically do not get the same advice and experience at these stores as you find at a specialized sports or outdoor store. The old adage "you get what you pay for" holds true with hiking boots, and a $25 no-name brand may not last long or be reliable on the trails. Luckily, sales are common at most outdoor stores, and great deals are not unusual.

It is best to shop for your boots at the end of the day, when your feet are tired, hot and swollen—a similar scenario to hiking. At the store, inspect the boots carefully and think about your needs. Are the boots waterproof? Does the stitching look sturdy? Are they too light or too heavy for your purposes? Do the sole and tread look rugged enough for the types of trails you plan to hike? Is the boot's material

GORP

too stiff or too flimsy, depending on how you plan to use them?

After you try on the boots, wearing the same type of sock you plan to use while hiking, walk around the store and, if possible, up and down stairs and any incline or decline. Some stores even offer mock trails and other apparatus to mimic trail conditions. If your toes feel scrunched, your feet slide around in the boot, or your heel lifts, then you probably have a poor fit. You should also consider whether the boots provide enough ankle support or perhaps too much.

Although it is normal for new boots to be somewhat stiff, keep in mind that any discomfort will become magnified once you are tramping around on the trails. The boots should be close-fitting yet comfortable, with room to move your toes and instep.

After you make your purchase, it is time to break in your hiking boots by wearing them around the house and then on short walks. The first time you take them on a longer jaunt, don't be surprised if you discover a few hot spots or blisters; however, this should disappear after a few outings.

But that day on the mountain, when my ankles throbbed and blisters decorated the soles of my feet, my hiking boots were beyond breaking in--they were simply broken: old, worn and flimsy. Hobbling down the trail the next day, following the sensible boots of my husband, I decided to follow his unspoken example. No more told-you-so looks for me; I had learned my lesson.
–Erin Fanning

Make sure to enter our annual
Creative Contest!
Entry deadline is September 1, 2006!

Visit us at www.WhisperintheWoods.com for even more ways to Discover, Explore and Appreciate the natural world.

GORP

Survival Skills
the rule of three's: shelter before food

How does the outdoors person prioritize survival needs when the unexpected happens? Making quick rational decisions can be the difference between life and death.

The rule of three's is a guide to help prioritize immediate survival needs. It goes like this: A person has three minutes without oxygen; three hours without shelter; three days without water; three weeks without food, and three months without human companionship. Some of these time frames can be debated; please remember these are just averages and are used as a guide.

Some of these three's are self-explanatory—like the need for oxygen—while others should be considered more closely. Let's look at the need for shelter versus food.

Hunger tends to lead us to believe that food is more important than shelter, when in reality humans are able to survive weeks without food. Most people forget that hypothermia can strike even when temperatures are in the low fifties.

If you should find yourself lost, construct a simple, A-frame shelter using a survival blanket, small tarp or plastic sheet. String a line between two trees, throw the tarp over top, and stake it down. If you don't have anything with you, make a shelter out of leaves and sticks. Insulate yourself from the ground and get out of the elements. In the morning, you can take care of your next priority.

Using the rule of threes can help lead the outdoors person to make the right choice in a survival situation. Making up your own survival scenarios is a great way to train the mind. Use different environmental conditions, varying equipment types and challenges.

Being mentally prepared before heading outdoors can help you make critical survival decisions when the unexpected happens.
–Shawn Grose

There can be no happiness if the things we believe in are different than the things we do.
-Freya Stark

Taking The Road Less Traveled...

The Chosen Path, Fall - Photography by Stacy Niedzwiecki ©2005

The 2007 HUMMER H3
From $29,945*
Plus tax, title, lic. & doc. fees

This page is brought to you by **Harvey HUMMER**
Our all new showroom is now open at 2500 28th Street SE
in Grand Rapids! (616) 949-1145 or 1-866-514-HUMMER
www.harveyhummer.com

Isle Royale Wolves

the synergistic ebb and flow of Isle Royale wolves and moose

by Jan Ferris

Following a day of backpacking at Isle Royale National Park, I snuggled in my sleeping bag. Echoing loon calls from Moskey Basin lulled me to sleep.

But there was another animal call, drifting down from the ridge top, longer and lower than the wail of the loon. I stopped breathing momentarily and listened, then popped my head out of my warm cocoon when I was sure I was hearing correctly.

After twelve years of visiting Isle Royale, I was finally hearing a wolf howl. More chimed in, each with its own pitch and deliberate song. The chorus drifted through the campground and echoed across Moskey Basin, making the hair on the back of my neck stand up.

Some hikers unfamiliar with north woods wildlife mistake the wail of the loon for wolf howls. But once you hear them together, there is no mistaking the wolf. Maybe it's just wishful thinking, as the goal of most visitors is to see moose and hear wolves. A good percentage of those who venture into the backcountry are lucky enough to accomplish both.

One cannot tell the story of gray wolves on Isle Royale without talking about moose, because their existence is intertwined. It is believed that moose first swam across Lake Superior from Minnesota and Canada in the early 1900s. With no competition for food from deer or caribou, moose found a plentiful supply of aspen and balsam fir. The population grew to about 3,000 by 1930. There were also no predators to thin the herd. But when the food supply was exhausted, numbers dropped to 400 or 500 by 1935. Lucky for the moose, a forest fire burned 20 percent of the island in 1936, and resulting new aspen growth made good habitat for almost three decades, so numbers grew again.

But something happened during winter 1948 that changed life as the moose knew it. An unknown number of wolves—perhaps just a pair—crossed an ice bridge from Canada and liked what they saw. They found plenty of old moose, their bread and butter to sustain them, and began reproducing. In an aerial survey in 1959, a pack of fifteen wolves hunting moose was captured on film for the first time. One lone follower and a couple others were also spotted. This was the beginning of the life and death saga that would reveal much about both species to scientists around the world.

Even before Isle Royale National Park was dedicated in 1946, scientists found it to be the perfect natural laboratory. In 1958, Durward Allen of Purdue University initiated the world-renowned wolf-moose research project there.

Moose and wolves share the 210 square miles with thirteen other species that made it there. Since it is a national park, no hunting is allowed. The park is only open from May to October, and each year only about 17,000 visitors come, giving moose and wolves plenty of space and time without interference from humans.

Now in its forty-eighth year, the Isle Royale wolf-moose relationship remains the longest-studied population of vertebrates, and possibly that of any organism, ever. Rolf Peterson took over the project in 1970.

cycle of both, but it wasn't until I read Rolf Peterson's book *The Wolves of Isle Royale, A Broken Balance*, that I really began to see the effect wolves have on moose, and vice versa.

Peterson and his assistant John Vucetich do a winter survey with pilot Don E. Glaser. They observe wolf-moose behavior from the air, photographing brutal life-and-death struggles. They also count them and note kill sites.

During summer, Peterson, his wife Candy, and Vucetich lead volunteers who pay for the privilege to retrieve bones of dead wolves and moose so researchers can perform tests. They also monitor wolves that have been fitted with radio collars, two at the present time.

As in any predator-prey relationship, populations rise or fall in tandem. As the moose population ages, wolves find weakened individu-

...some hikers unfamiliar with wildlife mistake the wail of the loon for wolf howls...

He has become a well-known and well-respected expert on moose and wolves, as well as a spokesman for the future of both species on Isle Royale. The wolf-moose study is funded by the National Park Service, National Science Foundation, Earthwatch, Michigan Technological University, and individual donors.

Since I began visiting Isle Royale in 1984, I have followed the predator-prey relationship of the moose and wolf through articles. I have always been aware of the feast and famine

als easier to hunt, providing more food for the pack and increasing the survival rate of pups. But with more wolves, moose populations decline as easy-to-hunt calves and older individuals are culled. The surviving healthy moose are more challenging to hunt, so wolf populations decline. As more moose survive and reproduce, wolf populations increase again, and the cycle repeats itself.

According to Peterson, the average moose life span is twelve years. The oldest bull recovered

Isle Royale

General park information is available from Isle Royale National Park, 800 East Lakeshore Drive, Houghton, MI 49931-1869 (906) 482-0984 or www.nps.gov/isro

at Isle Royale was seventeen years old, and the oldest cow was twenty. Moose are susceptible to periodontal disease, arthritis and osteoporosis. Moose with arthritis in the lower back and hip joints quickly fall victim to wolves.

Wolves commonly suffer from broken ribs and other bones due to swift kicks from strong moose. Old wolves' teeth become too worn to kill and eat meat, causing many of them to starve. Wolves that trespass into another pack's territory are often killed, commonly lone wolves or alpha males and females. The oldest known wolf at Isle Royale lived to be twelve.

In 1980, a new factor was introduced to this tightly woven relationship when canine parvovirus reached the island, either from a visitor bringing an infected dog or by being carried on a hiker's boots. Peterson describes canine parvovirus as an acute viral disease that causes severe diarrhea and dehydration, often followed by death within a couple of days. Despite successful annual reproduction, Peterson estimated at least fifty-two wolf deaths between 1980 and 1982. The population rebounded briefly only to crash again, and by 1988 only twelve wolves remained. The canine parvovirus disappeared just as quickly as it appeared, however, and the wolves slowly regained ground.

While wolf numbers were down, the moose population skyrocketed, with a high of 2,400 individuals in 1995. As a result, moose were starving due to over-browsing and a lack of fire to encourage forest regeneration. They were also infested with winter ticks, or moose ticks. These parasites, 10,000 to 30,000 per moose, weaken them by sucking their blood, inducing malnutrition, and causing young moose to suffer hypothermia as they rub off their hair due to irritation.

Wolves took advantage of the increase in moose and their weaknesses, and wolf numbers rebounded. But a new problem was discovered: the wolves were severely inbred, and they had lost roughly fifty percent of the genetic variability of mainland wolves. Scientists feared that the wolves would not be able to reproduce, and they were faced with the decision of whether to bring new wolves to the island or let nature take its course.

Weather conditions also play a role. Several sequential winters of deep snow made browsing especially difficult for moose and hunting easier for wolves, so that by 1997 only five hundred moose and twenty-four wolves remained. Global warming is also blamed for low moose numbers, since moose cannot tolerate hot weather, do not sweat, and high temperatures alter their feeding behavior. The warming trend has left Isle Royale without an ice bridge to the mainland during most winters, making it less likely that more wolves will ever cross over again.

The current population is estimated at 450 moose and thirty wolves, a ratio of only fifteen to one. The study found that the average long-term kill rate is about two moose per wolf per 100 days. The wolves are dispersed in three main packs: Chippewa Harbor Pack with eight members, Middle Pack II with thirteen, and East Pack III with nine.

With the present wolf population being so high, you are more likely to hear howling wolves anywhere on Isle Royale, but especially at the more remote central campgrounds. But scientists and park staff are doing all they can to ensure public safety and to discourage wolf-human contact. Wolf sightings should be reported to a park ranger, and you should keep your distance and not interfere with their natural behavior.

It is amazing to me that either species is still alive. In his book, Peterson ponders the future of wolves and moose on Isle Royale: "One must realize that neither of these species were 'native' to Isle Royale, so it is debatable as to how far we should go to manage them, and when we should step in to 'help' them," he says. "Should we favor predation or species extinction?"

One can only hope that the current situation will remedy itself naturally. When questioning whether scientists should step in to "prop up" either species, Peterson offers this food for thought: "The wolf reduction of the 1980s illustrated how wolf predation had previously kept moose in check and allowed the forest to grow. In the complete absence of wolves, moose might so destroy their own resource base that they would face extinction themselves."

If either species ever risks extinction on Isle Royale, we must trust that scientists and park staff have learned enough to make wise choices. In the meantime, we can be thrilled by the howl of the wolf and the sight of moose on Isle Royale, and understand just how special these opportunities are.

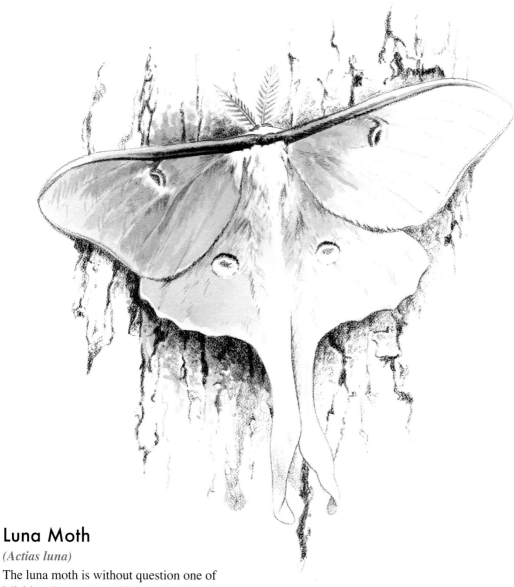

Luna Moth
(Actias luna)

The luna moth is without question one of Michigan's most beautiful moths. Its pale green wings are up to four inches across, with long, curved tails on the hind wings. Although the range of this moth is all over eastern North America from southern Canada to Mexico, in Michigan it is most often found in the UP, and of course, on Isle Royale. The favorite food for the caterpillars in this area is birch, while their southern cousins feast on walnut, hickory and sweet gum.

Striped Black Fly
(Simulium vittatum)

If wolves and moose are what Isle Royale visitors most want to see, blackflies are the critter they most want to avoid! These stocky, hump-backed flies pack an incredible bite. Larvae live in fast-moving streams with high oxygen content. They literally hang on for dear life, producing a string of silk from their mouths to attach to a rock or branch, and tiny hooks on their abdomen hang onto the silken lifeline. Female blackflies are blood feeders, and are most active during full daylight hours in late spring and early summer.

Mink Frog
(Rana septentrionalis)

Distinguishing a mink frog from its close cousin the green frog is not easy until you hear them sing, or make really close contact. Minks have a low-pitched, repetitive call that sounds like distant hammering, versus the single-note "plunk" of the green frog. If you pick one up, you'll know which you have immediately, as the mink frog is named for the musky odor it releases when attacked! Minks are a bit smaller, topping out at three inches in length, versus four and one-half for the green. Their skin color is also a bit different, with a heavier pattern of darker blotches that do not form stripes on the hind legs like those of the green frog.

Goldeneye
(Bucephala clangula)

The goldeneye is a common migrant duck in Michigan, but Isle Royale is one of the few places to see this handsome diver with its young. For birders who rely on the white cheek patch of the male for identification, the brown, unmarked head of the female might be a challenge at first. During the thirty days of egg incubation, the male guards a small feeding area for the female, then moves on to molt while she rears the young. Look for these bobbing families in open water around Isle Royale in June.

Thimbleberry
(Rubus parviflorus)

Thimbleberry is in the same genus as raspberry. Luckily for berry pickers it is one of only two members of the bramble family that is thorn-free. It is a western species, but has eastern populations in the Black Hills and northern Great Lakes regions. In the UP and on Isle Royale, it is common at woods' edge, in clearings, and in young, thin woods. The large, three-lobed leaf is easily mistaken for an extra-large maple leaf. The bright red fruit is the plant's namesake, resembling a flattened thimble. Although it is very popular in jellies and jams, it is a bit tart "straight up."

Moose

(Alces alces)

Encountering an animal that is ten feet tall at the shoulder and weighs well over 1,000 pounds is not an experience you are likely to forget. Such an encounter is very likely on Isle Royale, as some of the camp sites are located on streams frequented by feeding moose. Don't take the encounter lightly, though—remember moose have poor vision, and cows are quick to defend a calf that you might not have even spotted. They regularly feed right in streams, sometimes submerging themselves to feast on aquatic plants and keep blackflies at bay as well!

by Janea Little
Artwork by Rod Lawrence

Island Spirit

a gray wolf's howl conjures an unanticipated image

Lasting Impressions

by Mark S. Carlson

There are voices that speak only when the wind moves through the shoreline pines.

There is a sweetness smelled only when the raindrops fall through the spruce boughs.

There are happy songs heard only when the warming sun shines over the balsam woods

…and there is a sacredness felt only when the loon calls from moonlit waters.

These things, these places, live as the soul of this island.

You cannot see these things until you see them with the heart, then they will be yours forever.

The songbirds' symphony for dawn was well underway as the painted hues of yellow and orange grew more heavily stroked with approaching sunrise. Cool breezes off the big lake carried the refreshing scent of spruce and balsam, while a loud rhythmic tapping sound echoed throughout my neck of the woods. It was a nearby pileated woodpecker searching for breakfast around the base of a hollow birch log. For a time, wood chips were flying in every direction faster than in a *Woody Woodpecker* cartoon. All of the elements making up this scene were comfortably soaking into my consciousness like the pages of a well-worn book by Sigurd F. Olson.

Joyously, I continued illustrating the story by making photographs of the silhouetted lakeshore as the sun peeked over Superior's vast, island-dotted horizon. With gently lapping waves ever-present in the background, I found myself being lulled deeper and deeper into the timeless moment. Suddenly, out of the distance came a subtle, yet harkening call. My pounding heart was the only sound I made as the hair on my arms and on the back of my neck rose to the occasion. My primal self was responding to this ancient voice of wildness, instinctively honoring the same raw energy that has tingled the souls of humans ever since their inception upon this planet. Two mournful howls from a gray wolf became the exclamation point to an already captivating first paragraph in this developing story about a day in the life of Isle Royale National Park.

As I tucked my camera gear back into my backpack and collapsed my tripod legs, my elation over my morning shoot this day was palpable. While walking along the craggy coastline trail, my mind couldn't help but replay over and over all the things that had enthralled my senses. With such a grand start to the day, I anxiously imagined even more wondrous events unfolding throughout its course.

Stopping frequently to ponder and photograph young trees emerging out of rocky crevices or to visually explore the colorful textures of map lichens that adorned the cliff faces, my adventurous day passed quickly into its closing chapters. It was then that I found myself, as many nature photographers can attest, asking the age-old question, "Where did the day go?" I consciously realized I was already shooting late light when the photographer in me began nudging

natural instinct never fail to pleasantly surprise me. I've learned to anticipate and enjoy this process of making photographs very much.

I knew the full moon would appear over the southeastern horizon of Lake Superior during or around the time of sunset. Even though it had been clear all day long, I began to notice a thick cloud bank, camouflaged in blue, hanging right above the water line as the sun was setting behind me. I was worried it would build and seriously jeopardize my opportunity to make any photograph whatsoever, causing twilight to quit illuminating the foreground rock formation before the rising moon could appear. Tension grew, along with sunset's waning afterglow.

As if on cue, an arching sliver of white light began to appear behind the quickly dissipating cloud bank.

...cool breezes off the big lake carried the refreshing scent of spruce and balsam...

my brain to remember a previously noted importance to this day. I finally recalled the astronomical fact and exclaimed, "No wonder it seemed so magical, the full moon is scheduled to appear this evening." Perhaps the howling wolf knew it all along.

Sometimes photographers *do* plan the photograph, but more often than not, in my case anyway, the photograph finds *me*. The best I can do is set myself up and be ready for the experience. Nature and

I hurriedly traversed the cobblestone shoreline to create the perspective of the full moon rising over the end of Scoville Point. Just as the lichen-covered rock and evergreens were being illuminated by twilight's last glimmer, I ended the day as I began it, photographing the ethereal spirit of this enchanted isle. Then suddenly, out of the distance came yet another subtle, harkening call…

Shimmering water

    dancing to a loon's love songs

        this wilderness moon

Full moon rising over Scoville Point, Isle Royale National Park

Nikon F3 camera with a Nikon 70-200mm lens. f/16 at one second on a tripod using Fujichrome Velvia 50 transparency film. Photo by Mark S. Carlson

To learn more about Mark's photography, including his Great Lakes photo tours, please visit:www.MarkSCarlson.com

Find the heating & cooling system in this photo

You're looking at it. With a WaterFurnace geothermal heating and cooling system, nature provides everything you need **for ultimate comfort.** A WaterFurnace system harnesses the natural and free energy found in the earth to provide year-round heating and cooling. And because it doesn't burn fossil fuels, WaterFurnace protects the environment instead of polluting it. **Comfort, reliability, energy-efficiency, lower utility bills.** WaterFurnace geothermal is always a natural fit. For more information, visit WaterFurnace.com or contact any of these local dealers.

① **Whitmore Lake**
Michigan Energy Services
(888) 339-7700

② **Lapeer**
Priority Service by Porter & Heckman
(810) 664-8576

③ **Dewitt**
S & J Heating
(517) 669-3705

④ **Big Rapids**
Stratz Heating
(231)796-3717

⑤ **Ionia**
Home Experts
(800) 457-4554

⑥ **Jackson**
Comfort 1 Heating
(517)764-1500

⑦ **Hart**
Adams Heating
(231) 573-2665

⑧ **Palms**
Lakeshore Improvements
(989)864-3833

⑨ **Cheboygan**
Jim's Handyman
(231) 627-7533

⑩ **Port Huron**
Ingell Refrigeration
(810)982-4226

waterfurnace.com | (800) GEO-SAVE

Dreaming with Wolves

a gray wolf imbues the camp site—
and the psyche—with equanimity

by Michael Rathbun

I've walked for two and one half days into one of the most remote wildlands left in the Lower 48: Isle Royale. Last winter I studied the map, read the guidebook; but this place—Chickenbone Lake—perhaps humorously named for its resemblance to a wishbone, demonstrates the limits of maps and books.

Not a splotch of blue ink on a piece of paper, not a few words on a printed page, Chickenbone Lake is where real, wild things are. Standing in the cedar swamp that drains into the lake's west end, this much is clear.

I walk out on planks laid by the National Park Service and stand still for 20 minutes, hoping the moose I hear splashing in the dark shallow water will reveal itself. I hear soft grunts and the slow breaking of alder branches, but the moose remains hidden. Long practiced in the art of standing still, the moose is unimpressed with my 20-minute effort.

Instead I'm simply scolded by the red squirrel that scurries about me. Down the planks, up a spruce, back down on the planks, the squirrel buzzes her questions. What brings

> *...sounds like canine paws slapping and digging in the dirt nearby...*

you here, and what do you bring here with you? How dare you, in fact! I nod to the squirrel, apologize for the intrusion, ask for her blessing. It's a few hours until dark, when the moose will become more active. Withdrawing to my camp, I yield the swamp back to the locals. *Pax vobiscum.*

Before me, Chickenbone Lake: my campsite at its west end, where the two extensions of the lake join in a sideways V. I cannot see the lake's northern arm. The southern arm extends away from me for about a mile. A few reeds are in the shallow margins. The lake is calm and quiet. I am seated on the ground at the water's edge.

Around me, the forest: white spruce, paper birch, balsam fir and aspen dominate the landscape for miles in all directions. Thimbleberry and alder compose the understory. A few yellow birch leaves drift out of the trees and fall roundabout. I perceive the ever greenness of the spruce, the curling white skin of the birch. The sky: blue and deep, as the sun declines. The earth: in places packed dirt, in places green grass like a lawn, which recalls a piece of Han-shan's poem:

> *Thin grass does for a mattress,*
> *The blue sky makes a good quilt.*
> *Happy with a stone underhead*
> *Let heaven and Earth go*
>   *about their changes.*

Changes—like the darkness that is falling. Now, in the dimmest of light, I perceive the dark silhouette of the cow moose. She forsakes her cedar swamp retreat and meditatively wades into the lake. She's one hundred yards to my left, and I hear her watery passage better than I see it—a slow, splashing stroll. Hurry seems not to be a word in the moose lexicon.

Changes—like the darkness that has fallen. I now lay in my down sleeping bag. The shelter I have carried, a lightweight plastic tarp, is suspended over and around me by pieces of cord. Open at both ends, the tent fills and deflates again and again with the passing night breezes, as if it were capturing the sighs of the earth and holding them there momentarily. A loon laughs on the dark lake. In time I am not awake.

A loud sniffing noise. What sounds like canine paws slapping and digging in the dirt nearby. I'm awake now and instantly experience the uncertain feeling that comes when an animal wanders into camp. I am prone, vulnerable. I cannot see.

I roll on my stomach and twist the head of a small flashlight. Its beam shoots from the tent's end, and a gray wolf, the size and color of a healthy sled dog, leaps into the light. For the longest of moments the wolf is perhaps five feet away. For the longest of moments I see the

wolf's left flank and hindquarters as it bounds away into the dark woods.

Instinctively, I desire the wolf to hear my voice, to emphasize that a human is present: "Hey! Hey! You're too close! You're much too close! You're *still* too close! Keep going!"

I roll over again, burrow into my sleeping bag. I listen. I invoke the names of many deities. I try not to leave anyone out. A friend in need is a friend indeed. My watch says it is 4:30 a.m. I listen.

A loud sniffing noise. What sounds like canine paws slapping and digging in the dirt nearby. Perhaps ten minutes have passed. On my stomach again. My flashlight again. Now my head is outside the tent. I scan to the right. A second wolf is there—smaller than the first, its fur more red, about twenty feet away. My voice again: "Hey! Come on, now! Shoo already!"

This wolf I do not surprise, I do not frighten. This wolf's eyes shine yellow in the flashlight's beam. This wolf stares like a state trooper standing beside a pulled-over car. Deadpan. We both know who's in charge here. We both know it isn't me. By the time I've decided I don't know what to do next, this wolf turns her head, melts into the surrounding blackness.

I lie on my back. I listen. Are there more wolves? I realize there is a major artery below my heart. I realize this because my clenched fists, pressed to my midsection, feel the pounding. My heart seems to think I am running the one-hundred-meter dash, or should be. Dear old heart.

It is 7 a.m., and there's enough light to see. Crawling from my tent, I survey the camp. My pack, with my food and possessions, is hanging in a spruce, unmolested. I walk over to where the smaller wolf had been sniffing and scratching.

A collapsible bucket I keep lake water in for filtering is there on the ground. It has been pulled over by its handle, dragged for a few feet. Its contents have spilled over the packed earth of my campsite in the shape of a fan. Where the water finished its run over the ground, I see the scratches, the claw marks in the dirt.

There are a dozen places where the ground has been scratched deeply by a paw with four claws. I place my hand over a typical mark and have to spread my fingers a little to match its width. I estimate the wolf had paws more than four inches wide. My, grandma, what big hands you have!

And what curiosity, even playfulness, too. The wolves, I realize, had never behaved threateningly. Their reputation was frightening, not their actions. I recall a few lines of Mahatma Gandhi: *One who is free from hatred requires no sword. A man with a stick suddenly came face to face with a lion and instinctively raised his weapon in self-defense. The man saw that he had only prated about fearlessness when there was none in him. That moment he dropped the stick and found himself free from all fear.*

Equanimity: a fine thing, but surely respect for the wild is in order. When in the presence of ultimately unpredictable animals, particularly animals that can pull down an aged moose, perhaps it is appropriate for the heart rate of a frail primate to increase. Perhaps our bodies possess a primal wisdom over which our philosophy is laid. Still, in the future, I resolve to be a more gracious camp host. Equanimity — maybe next time.

It is days later. My hike finished, I am drowsing in the passenger cabin of an Isle Royale ferry boat. The warm autumn sun beams through the plastic window. Lake Superior's waters pass beneath the boat with a rhythmic swish. Diesel engines drone hypnotically. Sprawled on a padded bench seat, like the seat on a school bus, I yield myself to sleep.

It is minutes later. A loping gray wolf comes near: head down, ears back, yellow eyes unblinking; the wolf comes very near, it seems to crouch. I jerk my legs in and awaken to the sound of my own startled voice — "Hey!"

I look around, embarrassed, but my fellow passengers are all out on deck. I am alone.

Or am I?

Artwork by Rod Lawrence

Everything in Nature has a Spirit

Brother Wolf

Jim Brandenburg

Early Snow

Once a Wolf

Monarchs Flock

White Wolf Leap

Wolves Tracks from Above

Underling Appeases Alpha

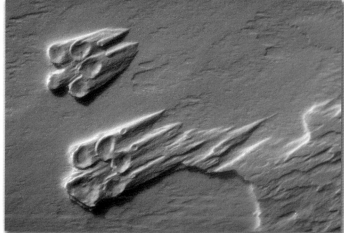
Wolf Tracks in Snow

Autumn 2006   31

Snowshoe Hare with Stick

Lily Pads at Dawn

Gayfeather in Fog

Snowy Owl

Pups Tug of War

White Wolf in Cotton Grass

Two Wolves Licking Rock

Ojibway Lake Island

Birch Grove

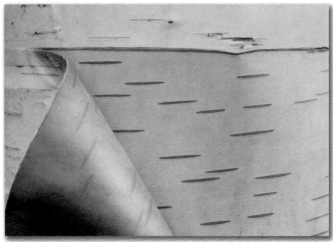

Birch

Autumn Wolf

Pond Ice

Prairie Smoke

Wolf Pup in Ferns

Lake Superior Beach

Wolf Skull in Snow

Quiet Pond

Fox

Watching for the Wolf

Bog Frost

Sing Back the Swans

Killing Frost

Birches with Soft Leaves

Day 10 Loons

Iridescent Raven

Arctic Wolf Pups

Photographer's Statement

One of the greatest joys I have in life is to know that through my photography, I might play a small role in increasing one's awareness of the natural environment and our dwindling species. *–Jim Brandenburg*

www.JimBrandenburg.com

Our environmental goal... leave no footprints.

At DTE Energy, minimizing our impact on the environment is one of our top priorities. That's why we let the US Fish & Wildlife Service manage habitat on more than 650 acres of our property. In fact, nine of our facilities are certified as official wildlife sites.

We've even planted more than 20 million trees since 1995. And we're working to provide efficient energy sources that are clean and renewable. It's all part of our commitment to enhancing the quality of life for today's society and future generations...step by step.

DTE Energy®

The Power of Your Community℠ | e = D T E®

Language of the Wolf

the howl of our fellow hunter beckons the spirit world

by Eric Alan

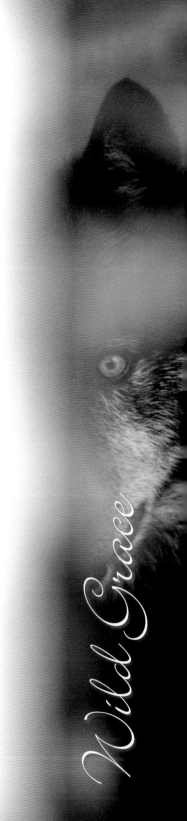

Wild Grace

Partially emerging from within the cocoon of a dream, I hear a howl cross the distance of the night. A wolf, I wonder? Then I fully awaken and realize: it's only a train horn, far down the valley. There are no wolves anymore in the mountains where I sleep. The primal feelings stirred in me by the noise were but an echo of instinct; a sound with enough similarity to call the feelings forth. In this particular place, the wolf is only noticeable by its absence. Its continued presence inside me is clear, though, evidenced by how easily its remembered cry pierces my dream cocoon.

Fortunately the wolf is not absent in all places. In recent years, the gray wolf has made a comeback in the Great Lakes states especially. Its resurgence has been touted as one of the great success stories of the Endangered Species Act, even though the wolf's comeback has been less successful in the Northeast and the West. The intensity of the ensuing fight over wolf reintroduction and its federal protection is just another measure of the animal's continued power in the human world—a power the wolves themselves have no way of knowing in our terms.

One glance at a wolf shows why it's been a potent symbol for the human race over the ages: it holds all the dignity, grace and beauty that a creature can carry. Still, the symbolism that persists in our current culture is generally of the wolf as wicked. As children, most of us were taught through fairy tales to fear them as frightening and evil creatures. Out of contact with wolves themselves, we learned their supposed ways from "Little Red Riding Hood" and "The Three Little Pigs." We learned to be afraid of the big bad wolf, and for most of us even now, to have the wolf at the door is to face a terrible threat (most likely financial). How deeply these concepts resonate, though we may speak the words with barely a concept of what's behind the cliché.

The wolf is truly a predator—of this there is no doubt. Countless ranchers can testify to the wolf's power to consume sheep and other delicacies; and in those lands where ranchers and wolves intersect, there is conflict and a need to balance the lives of both.

However, it was our own shifting relationship to the earth that made the wolf into an adversary. The wolf symbolism of previous cultures on this same land reflects this shift. The native societies of North America were largely hunting societies that did not raise livestock for wolves to eat. As a result, the wolf was regarded not as an enemy, but as a fellow hunter from which much could be learned. The stealth of the wolf—its stamina, its watchfulness, these and other qualities—were to be admired and emulated, not feared. In fact, so powerful is the wolf in many Native American beliefs that its howling is considered to be communication with the spirit world. For many Native American tribes, the wolf is variously seen as a teacher, a pathfinder, a messenger, a guide and a skilled traveler—but not as a sinister force. Even the early pioneer hunters of the American West told fireside tales of the wily lone wolves they encountered and battled or competed with. It was only when we turned from hunting to farming—and thus from wolf compatriot to keeper of wolf prey—that our images of the wolf shifted into more sinister visions. We created enmity where there was little before, or carried it to the new land from roots in agrarian Europe.

The wolves and their habitats were then slowly driven to the margins of existence. It's only been belatedly that we've awakened to realize that the victory was not so sweet—or a victory at all. The disappearance of the wolf means a disappearance of ecological balance that also harms us, a proliferation of prey that wolves controlled, and a disappearance of something more subtle but vital that cannot be measured by statistics.

What is it that's missing, when a howl in the night only turns out to be a train? What's lost when there are no wolf eyes to either learn from or fear? It's difficult to define, and it is as individual as our relationships with the land and the creatures with which we share it. I can only speak for myself: for me, the cry of the wolf touches a part of me that exists well below the layers of manners and language; below all that we call "civilization." It's a part of me that's still exceptionally wild and would rather die than be tamed. It's a part that knows joy and pain in a way far more raw and immediate—more alive and thrilling and excruciating—than any feeling city streets let through. To look into wolf eyes is to see a vision of humanity reflected back: a vision that shuns the alien concrete sprawl; sees it as some loud and mystifying illness. I see it through those wolf eyes, and I want to turn tail and run, too—to sit on the hills and howl until the spirit world hears. There are some things that can only be said in the language of the wolf.

Photo by Mark Graf
www.GrafPhoto.com

Tales in Bronze

Doe with Two Fawns

Roger Smith

This is the Life

Just Takin' a Break

Bonnie and Clyde

Bird Seed Bandit

Fawn meets Frog

Sleeping Fawn

Standing Fawn

Up and Running

High Alert

Artist's Statement

I was raised on a farm in southern Michigan, and I always loved the outdoors. I dreamed of being a wildlife artist as a teenager back in the 70's. Then life happened. Business, family, church and community consumed the next twenty some years.

My wife Vicki and I were on a drive home from Montana in the summer of 2000, when she said, "You should do some sculpting." I picked up the clay again with an eye toward bronze sculpture. I love the look and feel of sculpted clay and the permanence of bronze. I find real beauty in the organic shapes and curves of a Bison's back or a fawn's ear. Then I strive to share what I learn from His creation with those who view my art. *–Roger Smith*

www.rogerswildlifeart.com

enjoyNature.org

Travel Planner
nature travel destinations to explore in the Great Lakes region, whether you're hiking, paddling or cycling

Calendar of Events
plan your outings, from moonlit hikes to cross-country skiing to artist appearances

Seasonal Watch
when and where to find Nature in her finest splendor, from spring wildflower blooms to peak autumn color

Activities
Kids Treehouse, crafts and fun activities, Geocaching destinations

Marketplace
product reviews, from snowshoes to kayaks to backpacks

Galleries
post your own photographs, essays and poetry, or browse our featured artist galleries

E-Newsletter
sign up, it's free

Free E-Postcards
send beautiful images to your friends

... the next best thing to being there!

www.enjoyNature.org

Brought to you by *Whisper in the Woods* in partnership with numerous nature centers and non profit organizations in the Great Lakes region.

Scorn the Raven? Nevermore!

North America's largest songbird counts, uses tools and mimics human speech

by Jean Strelka

What they say about ravens—it's all true. Common ravens do eat other bird's chicks and eggs. They also sometimes damage farmers' crops. As a member of the family Corvidae, along with crows, jays, magpies and nutcrackers, these birds are often considered noisy, aggressive and undesirable. But there is also much about the common raven to respect and admire.

For one thing, the raven is adaptable. The common raven occupies a wide range of habitats nearly worldwide. Its raucous, deep baritone "brronk" may be heard deep in the forest, but it is just as likely to be encountered soaring over a desert, a sea coast, the mountains or rocky cliffs.

In the eastern United States, the common raven's range lies mostly in the northern tier of states, where, as the only species of raven, it could only be confused with its smaller cousin, the American crow. A common raven's length surpasses that of a crow's by more than six inches, and its wing span is fourteen inches longer, but size is often difficult to judge. When identifying a perched bird, concentrate on the head. The raven's bill is much longer and heavier than that of a crow, giving it a "Roman nose" appearance. Also, the raven's throat feathers are shaggy, rather than sleek, as on a crow. In flight, the raven's wedge-shaped tail is an obvious and reliable field mark. Different flight styles can also help separate ravens from crows; crows flap almost continually, while ravens often soar or alternately flap and soar.

The common raven is the largest passerine (song bird) in North America, and it is considered by many to be the smartest of all birds. Experiments have shown that ravens have the ability to count and to use tools. In a now-famous experiment, designed to test the raven's ability to solve a problem, biologist Bernd Heinrich dangled a piece of meat on a long string tied to a branch. Most of the ravens he tested came up with one solution or another for getting the meat. One clever raven pulled up the string in loops with one foot, holding the loops in place with the other until the meat was close enough to grab.

Communication is another of the raven's skills. More than thirty types of vocalizations have been noted, including alarm calls, territorial calls, comfort calls and chase calls. Ravens also mimic sounds in their environment, including human speech.

Besides being smart, ravens also provide a valuable service. They are an integral part of nature's cleanup crew. As scavengers, they will fly along roadsides in the early morning looking for roadkill. Lacking the hooked beak of a raptor, ravens must rely on other predators to break the skin of a carcass before they can feed. Ravens have been known to follow wolves and other predators to feast on their prey. Some researchers suggest that ravens may actually vocally alert predators to potential prey and then share in the bounty. In addition to carrion, ravens eat a wide variety of food, including live birds, reptiles and amphibians. They are also fond of eggs, insects, grains, fruit and garbage.

Fledgling ravens learn which items make good eating by exploring their world with their bills. During this exploratory phase of their lives, which lasts several months, the young birds pick up and manipulate almost all new objects they find, including sticks, stones, pine cones and leaves. They are also particularly fond of shiny objects. And although they soon learn that metallic items hold no food value, they often continue to play with them.

Young ravens also play alone or with other ravens. They often drop and catch objects, hang upside down by their feet, and play tug of war or king of the hill. Ravens have even been observed sliding down inclines. Although play is most common in young birds, adult ravens have been reported to play as well.

But while wheeling and tumbling through the air and doing spectacular dives may look like play, in some cases it might actually be part of the raven's courtship or pair bonding rituals. During courtship the pair will also soar together, sometimes with their wingtips touching.

It is believed that common ravens mate for life. And as life partners, the birds share domestic duties. While the female builds most of the nest, the male delivers sticks for her to use. And while she usually incubates the four to six eggs alone, the male brings her food and guards the nest. Both parents feed the young once they hatch.

Common ravens are devoted partners and parents; they clean up our roadside and possess intelligence on par with many of our pets. So next time you hear that deep baritone croak, scorn the raven—nevermore!

...the courting pair will soar together with their wingtips touching...

Common Raven
(Corvus corax)

Length: 24 inches
Wing span: 53 inches
Weight: 2.6 pounds

Adult male: Entirely glossy black with long pointed wings and a wedge-shaped tail, elongated throat feathers and a large chisel-like bill.
Adult female: Similar to male, but slightly smaller.
Juvenile: Similar to adults, but with dull brown tail and wings through their first winter.

Voice: Common call a deep baritone croak: "brronk." Other calls varied, including hoarse croaks and metallic "toks."
Behavior: Solitary or in flocks, may form large communal roosts. Flies with steady wing beats in long-distance travel; often soars or alternately flaps and soars, sometimes makes aerobatic rolls. Considered the most intelligent bird.
Mating and breeding: Monogamous, believed to mate for life. Both sexes participate in nest building; female incubates eggs, male feeds female and guards nest during incubation, and will sometimes incubate.

Nest: Location variable, usually coniferous trees or cliffs. Made mostly of sticks and lined with shredded bark or animal hair. Often repaired and reused.
Clutch: 4 to 6 greenish eggs marked with brown, about two inches in length.
Incubation: 18 to 21 days
Fledge: 38 to 44 days

Food: Mainly dead animals, but will also eat live birds, reptiles, amphibians, eggs, insects, grains, fruit and garbage.
Habitat: Highly variable
Territory: Unknown
Range: In the east, found mainly in the northern tier of states.

Birch

Sunlight brightened naked
white
birch reaches, touches
deep cobalt blue
sky

Denise R. Baker

Photo by Steve Brimm
www.Brimmages.com

Resource Guide

Beautiful Fall Colors-See them!
Canoeing • Hiking • Driving

Free Visitor's Guide
Oscoda Area Convention and Visitors Bureau
P.O. Box 572 • Oscoda, Michigan 48750
1-877-8 OSCODA - www.oscoda.com

 **Rock Harbor Lodge**
Isle Royale National Park Michigan

Isle Royale National Park
200 islands. 165 miles of trails.
Lodge and ferry packages.
Lakeside Rooms · Cabins · Dining

www.RockHarborLodge.com
(270) 773.2191 (Winter)
(906) 337.4993 (Summer)

Refuge at the Light Station

Stay at the adaptively restored 1923
U.S.C.G. Crews Quarters at the Great
Lakes Shipwreck Museum, Whitefish
Point, Michigan's Upper Peninsula

National Historic Site
(888) 492.3747
www.shipwreckmuseum.com

Outdoor Adventures
DOWN OUTLET

4144 M-72, Acme
Across from the Resort

(231) 938.9779
www.downoutlet.com

Travel Tip: Tread Lightly

Tread Lightly is a practice that allows you to enjoy our forests without changing or damaging them. It is a willingness to assume responsibility to care for the land and respect the rights of those you meet along the way and those who follow you. The five basic principles of the Tread Lightly program are to:
- Travel and recreate with minimum impact
- Respect the environment and the rights of others
- Educate yourself, plan and prepare before you go
- Allow for future use of the outdoors by leaving it better than you found it
- Discover the rewards of responsible recreation

Resource Guide

Pine River Homes

Builders of Distinctive
Log & Timberframe Homes

Our Homes and Customers are Individuals

(810) 367-7521
www.pineriverrlh.com

Hike the woods, *bike* scenic byways,
canoe or kayak pristine rivers, *birdwatch*
amid splendid fall foliage. Breathtaking
natural beauty in every season.
Northern Michigan Preserved.

Benzie County
(800) 882-5801 - www.visitbenzie.com

Complete Professional Real Estate Services
Suzy Voltz, Associate Broker

(231) 352.7123
408 Main Street
Frankfort, Michigan 49635
www.C21.SleepingBearRealty.com

Bug Baffler®
The Essential Bug Protection

Mosquito/Blackfly Barrier
Protective clothing for outdoor activities
Like wearing your own screenhouse
(800) 662.8411
e-mail: sales@bugbaffler.com

www.bugbaffler.com

Mark Graf Photography
*Artistic Interaction
with Nature*

Fine art prints and
stock photography available
Specializing in Michigan and
the Great Lakes region

(586) 215-2512

www.grafphoto.com

America's Finest Canoe Outfitters!

Dan & Cathy Waters, Owners

(800) 255.2922
111 East Sheridan Street
Ely, Minnesota 55731
www.canadianwaters.com

Resource Guide

Williamson Real Log Homes

Full service builder of log homes
Serving Michigan and Northern Ohio

Brian Williamson
(734) 847-5782
bwilliamson810@msn.com
www.RealLogHomes.com

Attract wild birds with mealworms!

Free Brochure on Request

(800) 318-2611   Fax (513) 738-4667
P.O. Box 188
Ross, Ohio 45061-0188
www.theNaturesWay.com

"I'm afraid, Sylvia, that his care will be very expensive. However, Medicaid protects the wife of a nursing home resident. You should talk to an attorney with experience in financing long-term care as soon as possible!"

John B. Payne, Attorney
Estate Planning • Medicaid
Probate • General Practice
(313) 563.4900
1800 Grindley Park, Suite 6
Dearborn, MI 48124
www.law-business.com

If your spouse is in a nursing home, planning can make the difference between maintaining your standard of living and living in poverty.

Punzel . . . a living folk tale

in Michigan's
Tidendal Woods
So. of Traverse City

(231) 263.7427

www.thelegendofpunzelspond.com
www.punzelscandinavian.com

Buying Tip: Buying Binoculars

First, you must determine how you want to use the binoculars. Do you want to go birding in all types of bird habitat? Or do you simply want a closer view of birds at your backyard feeder? There are three factors to consider: power, close focus and field of view.

Power is essentially magnification capability. Having a designation of eight, eight times magnified, is considered average.

Close focus is the shortest distance that you'll be able to see in focus. If you are interested in birding in woods or your backyard, you'll want binoculars that have a short close focus.

Field of view is the width of the actual view. If you plan to use the binoculars to scan large areas, consider binoculars with a large field of view.

The Gift of Nature

As a gift for your family or friends, or even as a gift to yourself, *Whisper in the Woods* is a promise of quiet sunrises and gentle breezes.

Call Toll-Free
to Subscribe
(866) 943-0153
Subscribe online
www.WhisperintheWoods.com

Subscribers will enjoy the added benefit of receiving a collector's art print yearly.

Back Issues Available
See what you've been missing.

Order Today!

Prices vary, please check our website

Whisper in the Woods
P.O. Box 1014
Traverse City, MI 49685
toll free (866) 943-0153
www.WhisperintheWoods.com